Fire Prevention Blueprint
*Seven Disciplines for Building Effective
Fire Prevention Organizations*

By: B. Aaron Johnson

Fire Prevention Blueprint

ISBN-13: 978-1987426557
ISBN-10: 198742655X

Copyright © 2018 Brian Aaron Johnson
All Rights Reserved
www.TheCodeCoach.com

Table of Contents

Introduction 5

Discipline 1: Know Your Community 7

Discipline 2: Have A Plan 13

Discipline 3: Enforce the Code 19

Discipline 4: Conduct Plan Review and Field Inspections 25

Discipline 5: Investigate Fire Incidents 33

Discipline 6: Educate the Public 37

Discipline 7: Be Adequately Staffed 45

Final Word 53

Annex: A Risk Assessment Model 59

Introduction

The two biggest challenges faced by fire departments and fire prevention organizations around the country are budgets and personnel; specifically, having enough funds and personnel to provide essential fire prevention services.

How can a community build a functioning fire prevention organization? How can the organization prove its value? What programs and features should be offered? Where are the organization's best efforts and dollars spent? How can the fire prevention organization, programs, staffing, and budget be justified? The seven disciplines presented in the *Fire Prevention Blueprint* will address these challenges and can serve as a manual for establishing, organizing, and managing the fire prevention functions of a fire department.

However, the key to success in these situations, as in any emergency situation, is to have a plan of action and start working on the plan. This book is that plan.

Whether you are taking over a fire prevention organization, you are creating a brand new organization, or if your existing organization needs to be restructured, the *Fire Prevention Blueprint* is your guide. Based on historical context, current needs, best practices, published standards, and successful fire prevention programs, this guide presents the seven disciplines that must be in place for fire prevention organization success. Following these disciplines will lead to an effective and efficient fire prevention organization.

Discipline can be defined as, "an organization's responsibility to provide the direction needed to satisfy the goals and objectives it has identified[1]." These seven disciplines are structured to start with identification and creation of the organization's goals and objectives, and then the practical implementation to accomplish those objectives.

---

[1] IFSTA. *Essentials of Firefighting*. Pearson, 2013.

The seven disciplines outlined in the *Fire Prevention Blueprint*, are:

Discipline #1: Know the community.
Discipline #2: Have a plan.
Discipline #3: Enforce the code.
Discipline #4: Are proactive with plan review and field inspections.
Discipline #5: Investigate fire incidents.
Discipline #6: Educate the public.
Discipline #7: Be adequately staffed.

The object of this guide is to provide the framework and basic blueprint. There is an abundance of resources available on each of these topics and their subtopics. This guide references many of these tools and resources. For ease of access, these can be accessed from the resource page at, www.FPOblueprint.com.

Chapter 1: Know Your Community

Successful fire prevention and life safety start with knowledge. This is a knowledge of the community served, its current state, its needs, its structures, its risks, and hazards. The knowledge obtained will be utilized to direct the priorities and strategies of the fire prevention organization and create effective and efficient fire prevention and life safety program.

Understanding the community and collecting the data needed to create a plan is a three-step process. These steps are information gathering, data analysis, and strategy development. This process of gathering information, analyzing the data, and developing a strategy is referred to as a community risk assessment (CRA).[2] The CRA process is an exercise in compiling data, from a variety of sources that will provide a picture of the community and its fire and life safety history in order to predict and prevent future incidents and loss. The CRA process will reveal trends, community needs, and exposure to risk.

To start the information gathering process, two questions must be answered. What information do we need to gather? How can this information be collected?

NFPA 1730, *Standard on Organization and Deployment of Fire Prevention Inspection and Code Enforcement, Plan Review, Investigation, and Public Education Operations*[3], outlines seven content areas that should be assessed.

1. Demographics
2. Geographic profile
3. Building stock
4. Fire experience profile
5. Incident response profile
6. Hazard assessment
7. Economic profile

---

[2] NFPA 1730, Chapter 5 outlines and defines the community risk assessment process. NFPA 1730:A.5.1 defines a community risk assessment as "a practical data gathering and analyzing exercise".

[3] www.nfpa.org/1730 - for free access to this standard

Demographics have long been used within the private sector to advertise and market products, conduct research, structure political messaging, and in many other areas to target specific segments of a population. Demographics utilize statistical data to describe the composition of a community's population. Demographic factors to be considered may include age, gender, socioeconomic background, income, home ownership, recreational activities, religious affiliation, ethnicity, culture, language, and customs.[4]

A knowledge of a community's geography can answer three critical questions for strategic planning: Where are the hazards? Where are the high risk areas? Where are the most valuable resources?
The geographic profile describes the physical features and attributes of a community and can be divided into two categories: physical characteristics and human characteristics.

Physical characteristics are those geographic elements that occur naturally or are naturally caused. Physical characteristics can include landforms and bodies of water, weather and climate, vegetation, and animal life.

Human characteristics are those geographic elements and features that are human-made or caused by human. These environmental modifications can include bodies of water, bridges, tunnels, roads, railways, buildings, and monuments.

Identifying the geographic concerns of a community is important for various planning, education, and mitigation activities, including awareness, evacuation, vegetation management, hardening of structures, and possible ordinance or inspection activities.

The building stock identifies the core areas, aging structures, high-fire or high-life safety occupancies, and historical or culturally significant buildings or structures. Inventory of a community's building stock describes the various occupancies, type and number of buildings, the hazards, and risks presented.

---

[4] For additional reports on demographics see this books resource page at www.FPOblueprint.com.

Each structure within a community can be assigned a risk category.[5]

- High Risk. Buildings having a history of frequent fires and a high potential for life or economic loss, or a building in which occupants must rely heavily on the building's fire protection features, or rely on staff assistance for evacuation.
- Moderate Risk. Buildings having a modest fire history and present only moderate potential for life or economic loss.
- Low Risk. Buildings having little to no history of fire with minimal potential for life or economic loss.
- Critical Infrastructure. Vital assets, systems, networks, or structures whose damage or destruction would have a debilitating effect on the community.

High risk occupancies may be buildings such as apartments, healthcare, detention, assembly, and educational facilities. *Moderate risk occupancies* can be ambulatory health care, walk-in clinics, and industrial buildings. Storage, mercantile, business, and office buildings could be considered *low risk occupancies*. *Critical infrastructure* facilities are buildings such as power plants, water treatment facilities, public safety buildings, and special structures unique to the community.

Each structure in each of these risk categories is going to have different priorities and objectives for fire protection and life safety. It is the job of the fire prevention organization to prioritize and create programs that address the unique fire hazards and safety risks of each.

Every fire prevention organization should create a fire experience profile for their community. Knowing the communities fire history can reveal trends, recurring incident types, and "problem" locations. The fire profile can be created by reviewing incident reports. The fire profile should include data on fire incidents, fire deaths and injuries, fire cause and economic impact.

In conjunction with the fire experience profile is the incident response profile. This is an accumulation of data on all other

---

[5] NFPA 1730:3.3.3

incident response types beside fire. This information can show the most common calls that a community or area may receive. These may include, medical calls, hazardous materials incidents, natural disaster, or general calls for service.

A hazard assessment demonstrates the various and specific hazards that exist within a community and the vulnerabilities they present. Based on experience, forecasting, and subject matter expertise, the hazard assessment will create a list of potential threats and hazards to the community. The hazard assessment can show how the threat and hazard will affect the community and determine what tasks, functions, training, or resources may be needed to prevent these from happening, or ensure readiness to respond to these incidents.

The economic profile describes those facilities, functions, and activities that are vital to the financial sustainability of the community. The fire prevention organization should have an in-depth knowledge of the processes, events, structures, and assets that, if lost or cancelled, would have the most detrimental financial impact or contribute the most devastating financial loss.

How can this information be gathered? The information needed for community knowledge can come from databases[6], fire department personnel, and community "spies."

Your fire department personnel are out and active in the community. They are walking through the community and its buildings. They are involved in the communities, groups, organizations, and clubs. They are talking to people and observing who and what is going on. The more you know about the community you protect, the better prepared you can be. It is in this "walking around" through the community that knowledge of new technology, new industrial processes, or new structures or building methods can be acquired.

The ancient military strategist, Sun Tzu, said, *"Thus, what enables the wise sovereign and the good general to strike and conquer, and*

---

[6] For additional reports on data in the fire service see this books resource page at www.FPOblueprint.com.

achieve things beyond the reach of ordinary men, is foreknowledge." A key to gaining foreknowledge in our communities is in the relationships that we build. Sun Tzu lists five kinds of spies that must be utilized in gathering and gaining "foreknowledge." These five types of spies are representative of the types of relationships that we must foster.[7]

Local spies. Foreknowledge of a community and its coming needs can be found by understanding the communities history. Building relationships with long-term residents and community leaders is essential. By understanding its history and the goals, a picture of the future of the community can be formed. From this picture, plans for future department needs can be determined.

Inward spies. When it comes to new technology, processes, or materials, the representatives and users must be consulted. We must understand that we do not know every detail about everything. We must turn to the experts in the technology, processes, or materials. These experts are passionate about their product and can tell you every nuance about it. It is the firefighter's job to apply the product knowledge to practical application of fire prevention or fire suppression.

Converted spies. The fire service must look outside of itself in order to adequately plan for the future. We must work collaboratively with other organizations (non-profit, law enforcement, engineering, mutual aid departments, etc.). What do these organizations see in the future? How does the fire department fit into their plan? How do they fit into the fire department plan?

Doomed spies. There are those individuals who seem to have given up. These are the ones who have seen the history, seen plans made (or not), and still experienced failure. These may be the disgruntled ones. In talking to these individuals, much can be gained by understanding their mindset, and what they have seen.

Surviving spies. In magazines, books, conferences, and classes, the fire service has a wealth of knowledge to draw from. By knowing the stories and lessons learned from those who have experienced

---

[7] Johnson, Aaron. *Sun Tzu and the Art of Fireground Leadership.* TheCodeCoach.com, 2016.

what we are experiencing or have been where we are planning to go, we can establish foreknowledge of what to expect.

After the data has been gathered, it must be analyzed and evaluated. Raw data is useless unless something is actively done with it, so the data must be put to work. Based on demographics, building stock, fire and incident history, and the other community knowledge factors, data can be utilized to further the goals and objectives of the fire prevention organization.

There are three primary reasons for reviewing data: (1) to gain insights into fire problems, (2) to improve resource allocation, and (3) to identify community education needs[8]. This data can allow organizations to:

- Identify at-risk structures and populations.
- Implement educational programs.
- Terminate ineffective programs or operations.
- Define the root cause of fire incidents.
- Predict future fire and emergency incidents.
- Improve response times and station locations.

Successful and effective fire prevention organizations know their community. They become intimately familiar with their community's demographics, economics, geographical features, fire experience, buildings, structures, and specific hazards. Through data analysis and evaluation, specific risks that a community is exposed to can be identified. From this data collection and analysis process, a fire protection and life safety strategy to reduce risk can be formulated.

---

[8] USFA Fire Data Analysis Handbook -
https://www.usfa.fema.gov/downloads/pdf/publications/fa-266.pdf

Chapter 2: Have A Plan

The fire service seems to have plans and procedures for everything. There are SOPs and SOGs that provide standard operating directives, there are fire pre-plans, incident action plans, and organizational strategic plans. However, there often is no specific fire prevention plan. What fire prevention organizations need is a clear, focused, and strategic fire prevention plan.

The fire prevention plan comes from knowing the community and its risks, hazards, fire and incident history, response types and call levels. From this data, a risk reduction plan can be created that focuses on a community's specific needs and seeks to eliminate or reduce those high-frequency events or incidents. This plan will show how much time and effort should be spent doing fire prevention tasks, such as public education, investigation, inspections, systems maintenance, or plans review. Programs and processes that are not making an impact or have no real practical application can be cut from the plan, and those energies focused on more critical tasks.

After you have defined your community needs and identified risks and hazards, a strategy for prevention and mitigation can be developed. The fire prevention plan[9] outlines the programs and strategies that will be utilized to reduce, mitigate, or eliminate the risks posed to the community. The fire prevention plan should also provide a strategy for the long-term growth and continued functionality of the whole fire prevention organization.

A strategy can be defined as, a long-term plan of action designed to achieve a particular goal, an elaborate and systematic plan of action, or a long-term plan of success to achieve an advantage. *Strategy* implies a deliberate long-term course of action, not a quick fix. Creating a fire prevention plan is a long-term goal that will require a clear, concise plan of action for successful implementation.

---

[9] The current fire service buzzword for this plan is "community risk reduction" plan or CRR. This term could be used here, but we also want to be clear on the idea that the fire prevention plan should also provide direction for the organization as a whole, and not be singularly focused on community risk reduction.

In his book, *Fire Strategies - Strategic Thinking*[10], Paul Bryant outlines five properties of a fire strategy.

1. To be specific to the unique set of fire-related parameters of the community profile.
2. To be a clear and concise document, despite the necessary and sometimes complex processes throughout its drafting.
3. To have the necessary detail to enable effective planning and design; yet, not inflexible to changing technologies or philosophies.
4. To have realistic and achievable goals.
5. A fire strategy is an organic and dynamic object. It should be modified and adjusted for it to remain true to its inherent goal.

In 1975 the American Insurance Association published "Special Interest Bulletin No. 5, *The Value and Purpose of Fire Department Inspections"*. This bulletin outlined 7 objectives for an inspection program. As you develop your strategy, use these objectives as a template for an effective program.

1. To obtain proper life safety conditions.
2. To keep fires from starting.
3. To keep fires from spreading.
4. To determine adequacy and maintenance of fire protection systems.
5. To preplan firefighting procedures.
6. To stimulate cooperation between owners, occupants, and fire departments.
7. To assure compliance with fire protection and life safety codes, standards, and regulations.

Proper life safety conditions can be obtained by evaluating the adequacy of exits, protecting the path of egress, making sure that building evacuation plans are current, and determining occupant loads of the space.

---

[10] Bryant, Paul. *Fire Strategies: Strategic Thinking.* CreateSpace Independent Publishing Platform, 2013.

Have A Plan

Fires can be prevented by monitoring the hazards associated with a facility or process. Many people in the workforce become complacent as they conduct their daily responsibilities without incident. Public education, therefore, becomes an essential component to keep fires from starting.

The general public passes through our buildings every day, mostly unaware of the life-saving features that surround them. Structural features such as enclosures, firewalls, fire partitions, and fire doors must be inspected and maintained to keep fires from spreading adequately.

There are three primary reasons that a fire sprinkler may fail. The top reason that sprinkler systems fail is due to a lack of maintaining operational status of the system; this can be followed up by inadequate or incomplete coverage of the fire area or hazard to be protected. The final reason a sprinkler may fail is inadequate performance of the system itself. Any prevention program or fire strategy should include components that are designed to determine the adequacy and maintenance of the fire protection systems.

The best way to ensure success when fighting a building fire, saving lives, and preserving property is to pre-plan firefighting procedures. Fire protection programs should provide a clear layout of the building, its systems, related hazards, and special procedures or requirements.

Fire prevention organizations should work closely with the public and establish a good relationship with the building owners and facility managers within their jurisdiction. If a client is seeking the services of a fire protection or life safety consultant, a major part of the proposal should include a clear plan that outlines how cooperation between owners, occupants, and fire departments will be achieved.

With the myriad codes, standards, and regulations that abound, a fire prevention program should educate, interpret, and enforce these requirements. With the constant submission of new code change proposals, and the creation of new products and fire protection methods, a skilled fire strategist will be knowledgeable enough to assure compliance with fire protection and life safety codes, standards, and regulations is met and maintained.

In addition to planning for the fire protection and life safety of the community, fire prevention organizations must also prepare for their own organizational success, longevity, and relevance. In his book, *Plan 4 It: The 4 Essential Master Plans for Every Church,* Tim Cool details the essential elements for creating a Master Plan. A Master Plan can be defined as a comprehensive study of a community, facility, or industry that describes the short, medium, and long-term development plans necessary to meet future demand. The three dimensions of master planning can be described as:

- A programmatic study of current and long-range initiatives, and what facilities and resources may assist in accomplishing those plans.
- A vision of the future, beginning with today's realities.
- A clear and intentional big-picture view of the organization's future based on the hopes, culture, DNA, and desires of the organization.

When initiating a master planning process, there are four "sub"-plans that need to be created. It is the combination of these four plans that make-up the completed master plan:

1. Organizational Master Plan.
2. Financial Master Plan.
3. Facility/Resource Master Plan.
4. Sustaining Master Plan.

Organizational Master Plan. This plan identifies *who* we are, *why* we do what we do, and *how* we do it. The first step in the master-planning process is to know your organization, and be clear on what it provides or should be providing. Seven critical questions can provide clarity and focus in this process.

1. What is the vision for our organization?
2. Who is our "target" audience/customer/client/people group?
3. What is our DNA as an organization?
4. How do we define "value" for our organization?

5. What is our "story," and how should it be communicated?
6. If space, resources, or finances were not issues, what programs or offerings would be started, provided, or expanded?
7. If we do not start, provide, or expand the above service, what impact will that have on our community/customer/client?

Apply these to your organization and current situation. Based on your answers to the above questions, is your organization where it needs to be? What needs to happen for it to get where it should be? What needs to happen so that the impact and value added will be improved?

Financial Master Plan. This plan allows you to determine the financial feasibility of your organization's short and long-term vision and goals. How will the master plan affect the budget? Is the master plan and goals set, financially attainable? What facilities, personnel, equipment, or resources will be needed to accomplish the master plan? How can we ensure that the finances are available for the plan objectives? Do we need to adjust the plan (for a more realistic objective) or adjust our current expenses?

Facility/Resource Master Plan. This step of the master-planning process can help to determine if your existing structures, resources, and facilities are compatible with your long-term goals and direction, or if changes will need to be made. Do we need new items for the master plan to work? Or can we utilize or re-purpose what we already have? Conduct an audit of your current resources. Do you have the tools, equipment, and facilities currently available to achieve the Master Plan objectives? Can current and future programs and service offerings be supported? What resources are needed to provide that support?

Sustaining Master Plan. To *sustain* is "to provide what is needed for something or someone to exist and continue to exist." What is your plan for ensuring the longevity of your organization, its resources, and its programs? How will the organization be sustained financially? Is the funding source or business model viable? How much will updates for program resources, hardware, or software cost? How will you sustain the personnel necessary to run the programs?

This process is essential for the continued health of your organization, its personnel, and the benefits it provides. A comprehensive master plan will put your organization on the right track to make the largest impact in the community you serve. Lack of a master plan will produce an organization that is just maintaining, and will eventually lead to failure due to lack of preparation for the organization, and will completely diminish the organization's impact in the world.

Chapter 3: Enforce the Code

Code enforcement is a key component in preventing fire and life safety incidents from occurring in existing structures. This can be a daunting task as it can be the most time consuming and require the largest commitment of personnel.

However, by breaking down this large process into smaller pieces, we can ensure that all occupancies are inspected at regular intervals. Based on a community's building stock, it can be determined which occupancies are at the greatest risk for fire and an inspection schedule can be created to address these risks.

Fire codes and standard do not directly address the frequency of existing building inspections. How often should existing buildings be inspected? Should all buildings be inspected with the same frequency? What structure should be inspected more frequently or less frequently? What determines inspection frequency?

All structures within a community can be identified as high, moderate, low risk or critical infrastructure. The higher the risk category, the more frequent and extensive the inspections should be.

NFPA 1730 provides only basic guidance on how to determine the inspection frequency of existing buildings. This standard establishes the minimum frequency of inspections[11], as follows:

High Risk	Inspected Annually
Moderate Risk	Inspected Biennially
Low Risk	Inspected Triennially
Critical Infrastructure	Inspected per AHJ

---

[11] NFPA 1730:6.7 [12] There have been some discussions on methods of establishing definitive risk categories for structures. A uniform method so that what one jurisdiction considers high-risk would not be considered moderate or low risk by others, but a consistent method to evaluate these risks. One method that the author uses is included in the annex section of this book.

High risk occupancies may be buildings such as apartments, healthcare, detention, assembly, and educational facilities. *Moderate risk occupancies* can be ambulatory health care, walk-in clinics, and industrial buildings. Storage, mercantile, business, and office buildings could be considered *low risk occupancies*. *Critical infrastructure* facilities are buildings such as power plants, water treatment facilities, public safety buildings, and special structures unique to the community.[12]

All the structures in the community will fall into one of these risk categories. The occupancy risk classification of each structure will be determined based on the knowledge of the community and the community risk assessment. Determining the number of occupancies in each category will reveal the amount of inspections that are required to be conducted annually, biennially, triennially, or even more frequently.

In his book, *Fireground Strategies, 2nd Edition*, Anthony Avillo outline an acronym for remembering all the pertinent information on a scene-size up. The acronym is, "COAL WAS WEALTH."

Though intended primarily for scene size-up, this can be a valuable tool for pre-planning and understanding the hazards and needs on a facility or within a community.

Construction - what is the construction type?
Occupancy - what is the occupancy type?
Area - how big is the structure (square footage/stories/area)?
Life hazard - what danger to occupants and responders is posed by the structure?
Water - is there a dedicated water supply? Where are the hydrants located?
Auxiliary systems - are fire sprinklers or fire alarm systems present and operational?
Street conditions - is the structure accessible or obstructed?
Weather - what potential impact does weather have on the structure or operations?

---

[12] There have been some discussions on methods of establishing definitive risk categories for structures. A uniform method so that what one jurisdiction considers high-risk would not be considered moderate or low risk by others, but a consistent method to evaluate these risks. One method that the author uses is included in the annex section of this book.

Exposures - what buildings or structures are nearby?
Apparatus & personnel - what is the required apparatus and number of personnel?
Location - what is the building address? How is it located on the property?
Time - is time of day a factor for emergency response operations?
Hazards - what hazardous materials or processes are located in the building?

By asking these questions, one can start to prepare for various emergencies that may occur, and become familiar with what types of operations *actually* occur within our areas of operation.

A less extensive assessment but maybe more efficient, is the **S.C.O.P.E.** acronym. While the above assessment process provides a detailed picture of a facility, sometimes an overhead, big picture view is all that is needed.

Utilizing the S.C.O.P.E. acronym, a facility's features and risk can be quickly assessed. The S.C.O.P.E. sheet provides a one-page overview of a particular building or structure.

Statement of activities
Provide a general narrative of the type of work and activities that are conducted within the structure or facility.
Construction type
Select the construction type as defined in NFPA 220.
Occupancy
Select the occupancy type and calculate the occupant load of the building.
Protection
This space identifies what fire protection and detection systems are in place.
Exposures
The section outlines what is located on the surrounding sides of the building being assessed that may be impacted by a fire. This should also take into account how a fire in exposure would affect the building being assessed.

Utilizing these processes, structures within a community can be properly categorized by risk. From this, the fire prevention organization can determine the number of inspections needed and the amount of time and personnel required to conduct code enforcement.

The simplified process for determining inspection frequency for existing occupancies may look like this:

Step 1. Conduct a community risk assessment.
Step 2. Classify the occupancy risk of each structure.
Step 3. Determine the number of inspections to be conducted annually.
Step 4. Determine the necessary staffing level needed to complete the inspections.

Effective code enforcement also requires those "soft skills" of dealing with people. Communication is a primary required skill in the fire protection/life safety industry. Persuasion is the most powerful tool that we can have in our communications toolbox. Persuasion plays a significant role in educating our community on why violations need to be corrected, showing the value of our work, and in the successful acceptance of our designs and ideas.

In the 1930s, Alan Monroe, a speech professor at Purdue, created an organizational pattern for creating persuasive speaking. His pattern is referred to as, *Monroe's Motivated Sequence Pattern.* By structuring your communications according to Monroe's method, you are enabled to lead individuals to see and take action on the issue at hand. Monroe's motivated sequence pattern requires five steps:

1. Attention
2. Need
3. Satisfaction
4. Visualization
5. Action

If properly employed, these five steps can lead to a persuaded audience regardless of the topic.

Attention. Gain the attention of the listener. This attracts the listener to what you are about to say. You want to create interest in the audience. A typical tool to be utilized here would be story, questions, a quote, or facts.

Need. Describes the problem and demonstrates a need for change in the current situation. This details what the problem is. Proof that a problem exists can be validated to the listener by following these steps:

- State the problem.
- Provide an example of the problem.
- Provide statistics/testimony that shows the seriousness of the problem.
- Show how the listener is directly affected by the problem.

Satisfaction. Presents the solution, providing sufficient information and evidence to allow the listener to understand how it accomplishes the goal. This answers the question, "How will you satisfy the need?" The following five-step order will accomplish this:

- State the solution.
- Explain how the solution will work.
- Show reasoning behind your solution.
- Show successful past implementation of the solution.
- Meet and respond to any objections.

Visualization. This describes the benefit of the applied solution to the listener. At this point, you would want to bring in your visual aid to better enable your audience to see what could be if the problem was solved.

Action. Tells the listener what they must do, right now, to solve the issue. Ensure that your action steps are clear, concise, and have a clear completion timeline.

As important as the technical skill of code enforcement is, the people's skills are equally as important. Utilizing the information in this chapter and adjusting your approach when dealing with people may be the first step that is needed toward achieving code compliance.

Chapter 4: Conduct Plan Review and Field Inspections

The plan review process can let a builder or property owner understand the feasibility and expected costs of their project. It also provides a preview of what the fire department can expect to be coming to their community. The plan review process reveals site access, water supply, construction features, and fire protection systems availability. Hazardous processes that take place within the structure, or hazardous materials stored on-site can also be discovered in the plan review phase.

Compliance with construction codes and installation standards is ensured through the field inspection activity. Systems are tested for functionality, and the structure and operational features are inspected throughout the process to culminate in the building owner receiving his final Certificate of Occupancy (CO). The CO signifies that compliance standards have been met, and the building is safe for occupancy.

For fire prevention organizations tasked with conducting plan reviews, NFPA 1730 lists nine plan review elements[13] that should be examined.

1. Fire Protection Environmental Impact (Feasibility Study). The feasibility study should examine such items as response times for fire/emergency services, communications capabilities, hydrants availability, and water main requirements. Any special considerations, and fire protection alternatives or equivalencies should be documented and reviewed.

2. Water Supply and Fire Flow. These should be conducted to ensure that the available water supply requirements can be met. If they cannot, other options should be considered and decided upon at this time.

3. Emergency Vehicle Access. This should be based on the largest piece of apparatus that the responding department may have to use.

---

[13] NFPA 1730:7.7

Driving surfaces, widths, overhead clearances, loads, turnarounds, and dead-ends should be considered.

4. Construction Building Plans. This element of plan review should be conducted to determine code compliance, occupancy classifications, construction type, required fire protection features, fire resistance ratings, interior finishes, and any special hazards or structures.

5. Certificate of Occupancy Inspections. These inspections are carried out throughout the project and can include all the trades (plumbing, electric, HVAC, etc.), and fire protection systems. These inspections ensure that what has been approved on the plans is what is being installed in the building.

6. Hazardous Materials and Processes. Any hazardous materials or processes should be reviewed for proper storage, handling, transfer, containment, emergency planning, and fire protection.

7. Fire Protection System Plans. These reviews confirm that required systems are in place, designed properly, and work for the structure. These systems include sprinklers, alarms, smoke control, fire pumps, hood systems, kitchen hoods, elevator recall, and similar items.

8. Fire and Life Safety Systems Field Acceptance Inspections. These final inspections are in place to visually witness the correct operation of the fire protection systems, and confirm that all systems are in place and functional in accordance with codes, standards, and approved plans.

9. Certificate of Occupancy (CO) issued. This is the main objective of any building project. After all work is completed, and all items are confirmed to be installed and functional per the approved plans, the Certificate of Occupancy can be issued, and the structure can be put into use.

In large structures with multiple systems, the best way to ensure compliant construction and systems reliability is through commissioning and integrated testing. This "new" function of the fire prevention organization serves as a critical link between the plan review process and the physical field inspections. Commissioning is

Conduct Plan Review and Field Inspections

covered by NFPA 3, *Standard for Commissioning of Fire and Life Safety Systems*. Integrated testing is covered by NFPA 4, *Standard for Integrated Fire Protection and Life Safety System Testing*.

Fire and Life Safety Commissioning (Cx) is defined as, "a systematic process that provides documented confirmation that fire and life safety systems function according to the intended design criteria set forth in the project documents and satisfy the owner's operational needs, including compliance with requirements of any applicable laws, regulations, codes, and standards requiring fire and life safety systems."[14]

Although NFPA 3 does not require any certification for commissioning agents, it does outline knowledge and skills that a commissioning agent should possess. The Fire Commissioning Agent (FCxA) is the person or entity who leads, plans, schedules, documents, and coordinates the fire protection and life safety commissioning team, implements the commissioning process, and ensures that integrated systems testing is appropriately conducted.

With this as the primary objective, an FCxA should possess the following:
- Thorough knowledge of the recommendations of NFPA 3 and general industry practices.
- Be capable of providing an objective and unbiased perspective.
- Advanced understanding of the installation, operation, and maintenance of systems to be installed.
- Ability to read and interpret drawings and specifications.
- Capable of analyzing and facilitating resolution of issues related to system failures.
- Clearly written and verbal communication, report writing, and conflict resolution skills.

Fire and life safety systems commissioning takes place in four phases: planning, design, construction, and occupancy. During the planning phase, the owner's project requirements are laid out and developed, the fire commissioning agent is selected, and the

---

[14] NFPA 3:3.3.3.1

commissioning team is put into place, the commissioning plan is created, all planning documents and regulatory codes are reviewed and analyzed, and the commissioning plan is put into action.

The Commissioning Plan provides the framework for the building projects commissioning process. The Commissioning Plan provides an overview of the project and outlines the process, system, and schedule to be followed. All the required commissioning reports, inspections, and documentation will be included as part of this plan. NFPA 3 recommends the following Commissioning Plan structure:

- *Introduction* - an overview of the plan.
- *Commissioning scope* - identifies which building components, structures, and systems will be subject to, and included in the commissioning process.
- *General project information* - overview of the project, focus on key information, expectations, and deliverables. This should include references and overviews of the OPR and the BOD.
- *Team contacts* - contact information of all commissioning team members.
- *Communications plan and protocols* - provide direction as to the projects organizational structure and communication channels to be utilized.
- *Commissioning process* - detailed explanation and outline of commissioning and project tasks to be completed for all phases of the process.
- *Commissioning documentation* - listing of all documentation that will be required and utilized throughout the process.
- *Commissioning schedule* - specifies the sequence of operations, and outlines the timeframe, dates, and duration of commissioning and testing events.

The OPR, or owner's project requirements, is the document which will form the basis of all design, construction, testing and operational needs, and will drive the decision-making process. This document should include such vital information as infrastructure requirements, occupancy use and classification, future expansion requirements, applicable codes and standards, and any other special needs or

specific requirements. This can be a dynamic document that should be updated as necessary throughout the four phases of the building lifecycle. The OPR should include:
- Infrastructure requirements (roads, utilities, etc.).
- Facility type, use, and dimensions.
- Occupancy classification, anticipated load, and expected operations.
- Future expansion requirements.
- Codes and standards that apply to the facility (local, state, national).
- Specific user/owner requirements.
- Training requirements.
- Warranty, operations, and maintenance information.
- Integrated system testing, installation, and maintenance requirements.
- Specific performance criteria that will be expected.
- Any "third-party" requirements.

The basis of design (BOD) is the focal point of the design phase. This is a driving document that should clearly show the concepts, ideas, decisions, codes, regulations, and standards required to meet the owner's project requirements. It is in the design phase that fire protection/life safety system drawings should be reviewed, commissioning procedures outlined and scheduled, and all documents verified to ensure that they comply with the BOD.

The decision-making process and an explanation of all systems and components are described narratively in the Basis of Design (BOD). A useful BOD will include the following components:
- Applicable codes, standards, laws, and regulations (NFPA, OSHA, ADA, ASHRAE, etc.).
- Building description.
- Fire protection/life safety system objectives and decisions.
- Alternative or performance-based design, means, and methods.
- Testing criteria.
- Equipment and tools required.

It is in the construction phase that all systems are delivered, installed, and tested. During this process, the fire protection and life safety commissioning team should closely monitor the construction process as they will be responsible for maintaining the commissioning schedule, ensuring that all materials and their installation are in accordance with the BOD, confirm that all work is being conducted by properly licensed and qualified professionals, performing all testing and inspections, and document all actions and any issues. The final action of the commissioning team in this phase is final acceptance testing and turning over all close-out documents to the facility owner.

At the end of the construction phase, before the occupancy phase, a closeout package should be delivered to the building owner. The closeout package should include the following documents:

- A compiled list of all deficiencies and resolutions.
- Operation and maintenance manuals.
- All test results, documentation, and certificates.
- Plans and drawings.
- Warranties and warranty information.
- Spare parts and supplier listings.
- Recommissioning plan.
- Sequence of operation.
- Software for systems should be installed and delivered.

The occupancy phase is the final stage of the commissioning process. It is at this point that all "loose ends" should be tied up, all final inspections conducted (and passed), all test and inspection reports completed, and system maintenance and product manuals turned over to the building owner. It is vital that the owner and other related personnel are adequately trained on the functions, operation, and maintenance procedures of the system. Every effort should be made to ensure that this training is complete and high quality, as education is a key component in continued effectiveness of any fire protection or life safety system.

Integrated fire systems are those fire protection and life safety systems that "are required to operate together as a whole to achieve overall fire protection and life safety objectives."[15] An example of an integrated system might be a fire alarm, fire sprinkler, elevator recall, and smoke control. When a fire is detected, each of these items has a specific code required function to perform. Integrated testing ensures that these systems all work flawlessly together. Integrated testing is to be completed as outlined in NFPA 4, *Standard for Integrated Fire Protection and Life Safety System Testing*. Integrated systems testing would be required and called out in the Basis of Design document, created in the design phase of the commissioning process. Integrated testing activities would take place during the construction phase of the commissioning process.

This integrated testing is to be supervised and managed by an Integrated Testing Agent (ITA). The ITA is responsible for planning, scheduling, documentation, coordination, and implementation of the integrated testing for all systems. The integrated testing plan provides guidance, direction, and time-frames for all systems personnel. This integrated test plan must include the following components:
- Verification of proper installation per design documents.
- List of each system that is installed and is to be tested.
- All documentation for each system (as required by that system's code or standard).
- List and contact information for all members of the integrated test team, their responsibilities, and a denotation of which individuals are required to be present for testing.
- List of all equipment required for testing.
- System input and output function matrix.
- Final system drawings and diagrams are to be listed and referenced in the testing plan and available on-site.
- Narrative description of test scenarios and procedures, and documentation and approvals for the AHJ.
- The extent of systems and system functions to be tested.
- Testing schedule.
- Future integrated systems test frequency.

---

[15] NFPA 4:3.3.25.4

A thorough and complete integrated testing plan will keep everyone moving toward the same goal. The plan will ensure that all items are properly tested, will work together, and all required documentation is accounted for.[16]

---

[16] Johnson, Aaron. "When Is Integrated Testing Required?" *The Code Coach*, 5 Feb. 2018, thecodecoach.blogspot.com/2018/02/when-is-integrated-testing-required.html.

Chapter 5: Investigate Fire Incidents

Conducting origin and cause fire investigations provide a whole different set of data than can be gained through inspections or enforcement. Based on investigations, new hazards can be identified, and incident causes can be tracked.

The information and data gained from conducting fire investigations are useful for:

- Improving public awareness and education.
- Implementing more aggressive inspections.
- Providing input into firefighting tactics and operations.
- Modifying regulatory requirements for buildings and products.
- Preventing or mitigating similar occurrences.

Fire origin and cause investigations can detect product defects, determine fire cause trends, and prevent arson and related crimes. The data collected from the investigation process can play an important role in community risk reduction. Origin and cause investigation can be a time consuming, and sometimes slow-moving, process. The investigation process includes on-scene time, research and data mining, interviews, report writing, and case preparation time.

For departments that are operating at minimum staffing levels, the use of company officers can considerably decrease the workload of the fire investigator and other fire prevention personnel. Company officers can be trained to conduct the preliminary investigation. If a more in-depth investigation is needed or a crime has been committed, then personnel assigned to this specific task can be called.

NFPA 1021, *Standard for Fire Officer Professional Qualifications* outlines the knowledge, skills, and abilities[17] required for a company officer to conduct preliminary investigations for origin and cause.

---

[17] for resources to assist in developing a department training program, visit the resource page at, www.FPOblueprint.com.

- Knowledge of arson methods, fire causes, fire behavior, and documentation of investigative procedures.
- Know when to delay overhaul operations.
- Ability to properly secure an incident scene.
- Ability to recognize and protect potential evidence from damage and destruction.

A more in-depth investigation process will have to follow the requirements of NFPA 921, with personnel certified to the standard of NFPA 1033. NFPA 921, *Guide for Fire and Explosion Investigations* is the standard for fire investigations and serves as a guide to provide clear direction on the conduct of fire investigations. Fire scene investigation methodology[18] can be conducted in six steps. These steps interconnect with the seven steps of the scientific method[19].

Step 1. Receive the assignment. The investigator is notified of a fire and requested to the scene. This links with the first step of the scientific method, "recognize the need."

Step 2. Prepare for the investigation. Prior planning is essential. An investigator should have a ready supply of the tools[20] that he will need for the investigation. This would include common hand tools, as well as evidence collection kits, photography and computer equipment, and all necessary forms and documents.

Step 3. Conduct the investigation. Investigations should be conducted systematically, utilizing the same system and routine for every investigation. Investigations should start from the exterior, or least burned area, and move inward or to the most damaged area. Steps two and three of the scientific method are to "define the problem" and "collect data." At the fire scene and through firefighter and witness interviews, determine and define the problem. What was seen, where was the seat of the fire, what was happening in and around before the incident? Begin to collect data, this is collected through photographs, interviews, and evidence collection.

---

[18] NFPA 921:4.4
[19] NFPA 921:4.3
[20] Hemmerling, Debra. "Suggested Contents for the Investigative Toolkit." *InterFIRE, A Site Dedicated to Improving Fire Investigation Worldwide.*, www.interfire.org/res_file/toolsall.asp.

Step 4. Collect and preserve evidence. There are strict guidelines for the collection, handling, and preservation of evidence. These should be followed. In the evidence collection process, great care should be taken not to overly disturb the scene or destroy other critical fire scene parts.

Step 5. Analyze the incident. This coincides with the fourth step of the scientific method, "analyze the data." Collect and compile all the known information - evidence collected, photographs, witness and firefighter statements, interviews, and other data. An analysis of this information should allow the investigator to eliminate potential fire causes.

Step 6. Form conclusions. From the analyzed data and evidence fire origin and cause can be determined. This is its own process that is included in the last three steps of the scientific method. Steps five, six, and seven are: develop a hypothesis, test the hypothesis, and select the final hypothesis. Eliminate the causes that are not probable; the potential causes should be worked through, one-by-one to be developed, and tested. As the potential causes fail the testing process, they can be eliminated. A conclusion can be formed when a potential cause passes the testing process.

Chapter 6: Educate the Public

Behavior only changes with education. By identifying root fire causes, at-risk populations, and hazard areas of the community, a public education agenda can be set. Whether the population is senior citizens, young children, a college town, or the workplace, there is a multitude of existing programs that can be used to educate and reduce risk effectively[21].

Businesses stay in business by offering solutions to known problems, or to issues their customers might not even know they have. As a fire service, we are pushing ourselves out of business by not providing solutions to a very real and well-known problem, the fire problem! We can save time and money by tapping into readily available resources that meet the critical need and address the 'fire problem.'

Public education efforts should focus on programs that are interactive, engaging, and provide maximum benefit to the community. Determining which programs offer the greatest value can be found by reviewing the data collected in the Community Risk Assessment (CRA). Interpreting the data and identifying the risks will focus your attention on the programs that are most needed. These ten steps will help guide your public education selection and efforts.

1. Collect the data. Data can be collected from a variety of sources and should include local population and census information, socio-economic indicators, fire department run reports, and local or national trends.

2. Compare the data. The collected data should then be analyzed to find trends and common, or frequently occurring incidents. These incidents can then be broken down by population data such as age group, socio-economic status, and geographical area of occurrence.

---

[21] for a list of educational programing resources, visit the resource page at, www.FPOblueprint.com.[22] Schwertly, Scott. *How to Be a Presentation God : Build, Design, and Deliver Presentations That Dominate*. Wiley., 2011.

3. Identify the risks. The risks that the data show will become the basis for your public education program. Public education efforts should be designed to reduce or mitigate these community risks.

4. Identify root causes. The public education program should address the actual cause of the problem, not just the symptoms. To get to the root cause will require more in-depth analysis of the identified risks.

5. Define goals and objectives. The best objectives are S.M.A.R.T. objectives:

S - specific, M - measurable, A - achievable, R - realistic, T - time-based

6. Develop strategic partners. Reach out to other public and private organizations in the community. They will have a shared interest in your program and may provide additional resources and/or funds.

7. Develop the program. Create the public education programming, elements, and deliverables. Before spending a large amount of time creating a program from scratch, explore the many ready-made resources that are available. Get the program started and out to the public, do not get stuck in planning and preparing mode!

8. Implement the program. Deliver the program. Don't worry about everything being perfect, just get your program to the audience that needs it. You can always make changes and tweaks as the program grows.

9. Evaluate the process and impact measures. Your program should be regularly evaluated to ensure that you are reaching your target audience, and the message you want to be conveyed is being received.

10. Modify as needed. Within a set time-frame, the program should be reviewed to determine its impact. If changes to the message, audience, or delivery are required, then make them.

Educate the Public

With the many public education options available, it is easy to go for the program that has the most funding, the best resources, or something the individual educator enjoys. Chasing programs can take much time, money, and resources spent on a program that still might not be solving the 'fire problem.'

The general goal of the fire prevention organization is to prevent the loss of life and property damage due to fire. Where NFPA 1730 provides guidance on *what* needs to be done to accomplish this goal, NFPA 1452 provides practical guidance on *how* this can be achieved.

The *Guide for Training Fire Service Personnel to Conduct Community Risk Reduction* provides direction for fire departments to design and implement the community risk reduction plan. A key component of effective risk reduction is face-to-face interaction with community members. This can be achieved through public events, fire station visits, and most effectively, home visits. Community risk reduction programs and fire crews involvement in them produce three distinct benefits.

Material distribution. Home visits, interaction, and direct contact with the public can provide an excellent opportunity to distribute and discuss fire prevention, life safety, and emergency preparedness literature. With the abundance of documents and materials available, make sure that the selected items and literature are directly tied to the community's risk reduction plan and goals. Fire department personnel should take advantage of these opportunities to answer questions and create conversations that promote risk reduction initiatives.

Support other programs. Personal interactions and home visits improve the public perception of the fire department and allow the promotion of additional fire protection and life safety programs. Based on the conditions or personnel observed, some programs that may be promoted include:
- Smoke alarm installation.
- CO detection and alarm installation.
- Radon dangers and awareness.
- Residential fire sprinklers.

- Fire escape planning.
- Drowning prevention.
- Senior citizen risks and fall prevention.
- Fire safety for children.

Continuity of CRR programs. Effective community risk reduction is an endless cycle of planning, implementation, and evaluation. Home visits and discussion with community members and groups can provide feedback on current programs, and data for future community needs. As these programs gain traction and their effectiveness is tracked and demonstrated, community support for the department and CRR will be enhanced.

The most important component of community risk reduction (CRR) is strategic contact with the public. A strategic contact consists much more than handing out stickers or plastic hats at the mall. The strategic contact is a contact made that meets the objectives of the community's CRR plan and is immediately beneficial to the person contacted. This can most effectively happen in fire department home visits.

The purpose of the home visit is to make residents and homeowners aware of any fire hazards that may be present in the home, and ensure that smoke and CO alarms are installed. These visits and reports should be considered confidential. It is not the intent of the home visit to "punish" the resident, or cause problems within their community.

Here are some general tips and items to be considered before leaving the firehouse and engaging in CRR activities.
- Dress professionally in a uniform that clearly identifies you with your department.
- Ensure that a full supply of resources and handouts are available.
- Work only in teams of at least two people.
- Remember the primary goal of the CRR "mission" is to eliminate hazards to life and property. Be able to articulate this to your community clearly.

Witness testimony, community events and appearances, public education, all these put fire prevention in the public eye. How many of us have witnessed some fire official completely blow it while speaking to the public, delivering a training class, or giving a presentation? How many seminars have we sat through where a prominent presenter on an interesting topic, completely bores us to death? How many times have we experienced a near-miss at death by PowerPoint?

Many fire officials seem to consider their public speaking responsibilities as secondary, something not to be worried with. John F. Kennedy once said, "The only reason to give a speech is to change the world." In the fire service, we have something to offer that no one else does. We are in the business of saving lives (whether through EMS/fire response/fire prevention). Scott Schwertly[22] says, "The reality for both audiences and speakers is that a presentation that does not move something - be it people, products, ideas, or values - is merely wasted time."

For more than 30 years, Ken Davis, author of *Secrets of Dynamic Communications: Prepare with Focus, Deliver with Clarity, Speak with Power*, has been teaching people the art and science of public speaking through his SCORRE Conferences. In, *Secrets of Dynamic Communications,* Davis shares the SCORRE process and educates the reader on how to deliver compelling and memorable presentations.

The SCORRE elements are:

S - Subject. The main topic to be discussed.

C - Central Theme. Narrows the subject and focus of the talk.

O - Objective. A single sentence that articulates the focus of the speech.

---

[22] Schwertly, Scott. *How to Be a Presentation God : Build, Design, and Deliver Presentations That Dominate.* Wiley., 2011.

R - Rationale. The logical content that leads the audience to your objective.

R - Resources. Add light, color, and clarity and keep the listener engaged.

E - Evaluation. Self-evaluation that keeps the speaker focused on the objective.

John Milton Gregory[23] first published his seven laws of teaching in 1884. His template for teaching promises "a clear and simple statement of the important factors governing the art of teaching." These laws can be applied to any education, speaking, or training event to improve the trainer or educator's delivery and program.

Law 1. Law of the Teacher -- *"The teacher must know that which he would teach."*

Simply put, know your material. Teach what you know, do not try to over-reach your knowledge and experience. Spend plenty of time studying the material, make it fresh every time you teach. Do not be afraid to use other books, and outside resources.

Law 2. Law of the Learner -- *"The learner must attend with interest to the material to be learned."*

There are two types of motivation, intrinsic and extrinsic. The extrinsically motivated learner is there because he was told to be, is required to be, needs the class for certification or promotion, or someone else is paying the price for them to attend.

Intrinsically motivated learners are in the class because they have a desire to know and learn what is being taught, they want to enhance their career knowledge, and they are genuinely interested in the material. No matter how great the instructor is, the extrinsically motivated student will not receive the lesson to be learned. Transversely, no matter how bad the instructor is, the intrinsically motivated student will still take away value from the presentation.

---

[23] Gregory, John Milton. *The Seven Laws of Teaching*. 1884.

Law 3. Law of the Language -- *"The language used in teaching must be common to teacher and learner."*

Use common language and terminology that is common to your audience. Also, keep in mind that everybody has a unique learning style. Some people learn best by reading and seeing and others by listening. The most commonly encountered learning style seems to be hands-on training. Knowing this, and knowing the audience, you can prepare a lesson that utilizes a common language and the most effective teaching style.

Law 4. Law of the Lesson -- *"The truth to be taught must be learned through truth already known."*

Apply new concepts to past experiences of the student. Allow the students to reframe the lesson in their own words.

Law 5. Law of the Teaching Process -- *"Excite and direct the self-activities of the pupil, and as a rule tell him nothing that he can learn himself."*

Find the point of contact between the student's current situation and application of the lesson being taught. Teach students to explore themes and ideas themselves. Teach them to ask, "What? Why? How? Where? When? By whom? What of it?"

Law 6. Law of the Learning Process -- *"The pupil must reproduce in his own mind the truth to be learned."*

Help students take on the role of an "investigator" and work with them to reach the desired conclusions, or lesson to be learned. Encourage students to pursue lifelong learning and a constant pursuit of excellence, knowledge, and truth.

Law 7. Law of Review and Application -- *"The completion, test, and confirmation of the work of teaching must be made by review and application."*

To know the effectiveness of the presentation or educational program, the class of students must be evaluated. The most common method of evaluation is through the use of Donald

Kirkpatrick's four levels of evaluation[24] -- reaction, learning, transfer, business results.

The reaction level of evaluation assesses the students' initial reaction to the course. This is most commonly conducted through the use of a course or instructor program evaluation form. From these documents, the instructor can gauge whether the students received value from the training, and can receive feedback on course elements to add or remove.

The next level of evaluation, learning, is conducted through an end-of-class examination. This level of evaluation is meant to assess the amount of information that the student has learned and retained.

The third level of evaluation is transfer. The transfer level of evaluation occurs between six weeks and six months from the end of the class. This is to assess how much of the material the student has retained from the classroom to practical application in the field.

The highest level of evaluation occurs between six months and two years after the training program. The 'business results,' or ROI, evaluation allows the instructor to assess the financial impact and return on investment of the training program. This should answer the questions of, "Is what we are doing making sense?" and "Is it cost-effective to continue this line of training?"

A departments education programs should provide value to the community. Educational programs should have a clear goal and seek clear results. The results being a decrease in fire activity or loss of life within the community, with those results being clear and measurable through the use of data

---

[24] Kirkpatrick, James D., and Wendy Kayser. Kirkpatrick. *Kirkpatricks Four Levels of Training Evaluation*. ATD Press, 2016.

Chapter 7: Be Adequately Staffed

The most significant challenges faced by fire prevention organizations is budget and personnel retention. Money and people. More specifically, it would seem, lack of funds to hire, train, and retain personnel needed to carry out fire prevention functions.

The key to achieving effectiveness and efficiency is time management. Utilizing Peter Drucker's[25] three-step process for time management we can effectively record, manage, and consolidate our time and the time of our staff members.

Time cannot be managed until it can first be found. The first step toward time management is to record, track, and log how your time is currently being spent. The best way to accomplish this is through the use of a daily log. Always carry a notebook. Document every task that you complete throughout the day. At the end of each day, review where your time went and prepare the next day's schedule to determine where you want your time to go. At the end of each week, compile updates and current status of important projects and issues. These activities will provide an opportunity to review and evaluate where and how your time is being spent. At the end of each year, your inspection data, numbers, and time can be utilized to conduct a staffing task analysis to determine what exactly is being done, how long it is taking to do, and if staffing levels are adequate.

After reviewing where our time is going, it must be managed. The best way to start managing your time is to diagnose and eliminate non-productive and wasteful activities. To determine if a task is non-productive, apply this three-part 'diagnostic exam.'

1. Does this activity need to be done at all? What would happen if it were never done again?
2. Can this activity be done by someone else?
3. Does this task waste other people's time?

---

[25] Drucker, Peter F. *The Effective Executive*. Harper Collins, 1966.

Identify and eliminate those tasks that only serve to waste time and produce no results. Only do the tasks that require you to do them, otherwise, delegate the task to others. Eliminate those tasks that waste others' time, or find a more productive way to accomplish the goal so that no one's time is wasted.

Finally, look at the time that you have and consolidate what is there. This is when you take the time available throughout the day, put that time together, and focus on specific task(s) completion. It is best if this time can be uninterrupted. Working in this manner is a more effective and efficient way of working than to jump from task to task, or working in spats of short time spans. For example, schedule all your plan reviews to be conducted on a certain day or portion of ("plan review day"), make one day your day for meetings, set aside a specific time to conduct inspections and stay within the geographical area.

Understanding what activities the fire prevention organization does, or should do, and how much time those tasks and activities should take is a critical component in determining and justifying budgets for staff and personnel. What is the standard time frame for common fire prevention tasks? How long should inspection activities, plan reviews, or investigations take? How can we know the number of man hours to assign to tasks? NFPA 1730, provides some basic guidance on required timeframes for existing inspections, plan review, fire investigations and public education activities.

Existing Inspections -

1. Determine the amount of buildings in each risk category.
2. Determine the total amount of buildings at each inspection frequency - annual, biennial, and triennial.
3. Determine the total number of buildings that need to be inspected annually.
4. Divide the total number of buildings that need to be inspected biennially into a 2-year rotation.
5. Divide the total number of buildings that need to be inspected triennially into a 3 year rotation.
6. You will now have the total number of inspections that must be completed every year.

Be Adequately Staffed

7. Determine the average time required to conduct an inspection. This determination should take into consideration: <u>travel time</u>, <u>inspection process</u>, <u>research and paperwork</u>, and <u>follow-up or re-inspections</u>.
8. You will now have the total man hours required.
9. Apply the five steps outlined above to determine the amount of personnel needed to complete this task.

Plan Review/Field Inspections -

1. Based on occupancy type and/or building complexity. Tables 7.6.2(a) and 7.6.2(b)[26] provide time requirements based on occupancy and construction type.
2. Based on number of sprinkler heads or alarm devices. Tables 7.6.3(a) and 7.6.3(b)[27] provide time requirements based on number of devices.
3. Field inspection time is double the plan review time.

Investigations -

1. Departments should establish standard operating guidelines for minimum amount of personnel required for a fire investigation. The utilization of Company Officers for initial investigations is strongly encouraged.
2. The time required to conduct a fire investigation should include: <u>on-scene time</u>, <u>travel time</u>, <u>report writing</u>, <u>research</u>, <u>follow-up</u>, <u>court appearance</u>, <u>preparation</u>, and <u>data entry</u>. A complete list of items is included in NFPA 1730, section 8.6.1.2.

Public Education -

1. Determine what programs are going to be offered.
2. Determine how many times the program will be offered.
3. Consider time required for each program. Time considerations should include: <u>research and development</u>

---

[26] see tables in, NFPA 1730
[27] see tables in, NFPA 1730

of the program, promotion of the program, delivery of the program, and follow-up activities.

NFPA 1730, *Standard on Organization and Deployment of Fire Prevention Inspection and Code Enforcement, Plan Review, Investigation, and Public Education Operations,* provides a five-step system[28] to aid the fire prevention organization in determining the minimum staffing levels needed to fulfill these essential fire prevention and life safety functions adequately.

Step 1: Outline all services provided by the fire prevention organization.
Start the staff needs analysis process by listing all services provided, functions performed, and activities conducted. This should be an exhaustive and comprehensive list. The list should include all activities, no matter how many times they are performed or how much time they take.

Step 2: Determine time demand for each task.
Determine the amount of time it takes to complete each task, function, and activity listed above. This should include all components such as, preparation time, scheduling, research, conducting the activity, report writing, and any follow-up. You will need to determine the time each of these take on an annual basis.

Step 3: Determine total personnel hours required to complete activities.
Add up the total amount of hours required for all tasks and activities. If your organization has many different tasks, programs, and functions, these can be divided into groups to simplify, or further analyze the total hours required.

Step 4: Calculate personnel total availability.
This formula will determine the number of hours that each employee will have available. This must account for holidays, vacation, sick, training, and other times that the employee will not be available for work.

---

[28] NFPA 1730, Annex C

Be Adequately Staffed

Step 5: Calculate the total number of personnel required to perform tasks.

To determine the total amount of full-time employees that are needed to perform all fire prevention functions and tasks, divide the total task hours by the total available work hours.

Total task hours / total available work hours = total personnel required

Fractional values can be rounded up or down. If the number is rounded up, it can provide reserve capacity and provide some 'cushion.' If the number is rounded down, this could result in overtime or increased workload for personnel.

When considering staffing and personnel, think about this from Seth Godin. In his post, "Reasons to Work"[29], Godin makes the point that money and pay are often the most emphasized, yet other factors actually play a much larger role in our workplace and career satisfaction. Some of the reasons listed include for the pleasure or calling of doing the work, for the impact it makes on the world, for the reputation that is built in the community, however, number two on the list, below money, is to be challenged.

Bill Hybels, leadership expert, says that people perform at their best when they are slightly over-challenged. We all fall into one of the three categories, under-challenged, appropriately challenged, or dangerously over-challenged. The under-challenged do not have enough interesting work to keep them engaged. They are not provided with enough work to do. Unable to find contentment or purpose in their work, the under-challenged usually leave organizations for a more challenging position.

The appropriately challenged usually have just the right amount of work and tasks to accomplish. However, they are not being stretched and are only maintaining what is currently in place. They are not advancing the organization or improving its service to the community. They are not creating.

---

[29] "Seth's Blog." *Reasons to work.*, 23 Nov. 2010,
sethgodin.typepad.com/seths_blog/2010/11/reasons-to-work.html.

The dangerously over-challenged are working themselves to death. Often, this comes at a high cost to their families, health, and general quality of life.

Most employees fall into the upper under challenged/lower appropriately challenged area. This results in employees that are largely unhappy with their work, merely going through the motions, and not producing at their highest level; their full potential to the community and organization is never realized.

Our best work is accomplished when we are working and functioning in the lower third of the dangerously over-challenged level. In this position, we are continuing present responsibilities while being stretched and encouraged to grow our organization and its impact in the community.

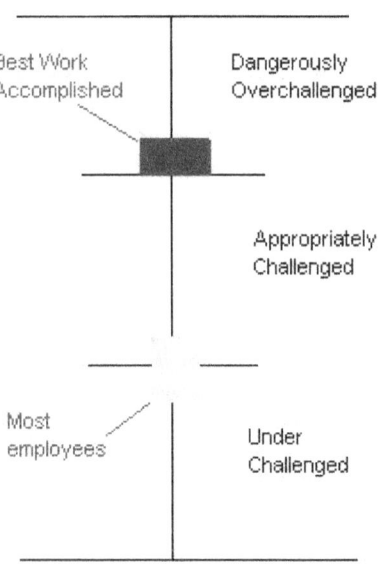

As a leader, it is part of your responsibility to ensure that your employees are adequately challenged. Do you know your employees well enough to determine what challenge level they are currently at, what level they are capable of, and what level they need to be stretched to? What do you need to do to increase (or decrease) their challenge level?

Final Word

In the year 1824, Edinburgh, Scotland was faced with a fire crisis. Major fires were occurring throughout the city, the insurance company fire brigades were less than effective as they lacked discipline and failed to work together. The municipal leaders, not happy with the situation, set out to take control of their fire problem. So it was that one of fire service history's most influential and progressive thinking officers, James Braidwood was selected to become the first Master of Engines for the Edinburgh Fire-Engine Establishment.

In 1830, Braidwood wrote, "Not having been able to find any work on fire engines in the English language, I have been led to publish the following remarks, in the hope of inducing others to give further information on the subject." These "remarks" became the 138-page book, *On the Construction of Fire Engines and Apparatus: The Training of Firemen, and the Method of Proceeding in Cases of Fire*. Braidwood's book covers much more than how to build a fire engine. It provided one of the earliest guides "on the causes of fires and the means of preventing them."

In recent years, standards such as NFPA 1730, NFPA 1452, NFPA 1300 have been created to discuss and present the "new" concept of community risk reduction or CRR. This story of James Braidwood from 1830 demonstrates that the principles of fire prevention, and effective fire prevention organizations have been around for a long time. Though the terminology may change, new buzz words may come and go, these seven disciplines for effective fire prevention organizations remain the same.

Effective fire prevention organizations:
1. Know their community.
2. Have a plan.
3. Enforce the code.
4. Are proactive with plan review and field inspections.
5. Investigate fire incidents.
6. Educate the public.
7. Are adequately staffed.

On *knowing your community* Braidwood wrote:

"...every exertion should be used to keep the firemen on good terms with the populace."

"He should also make himself well acquainted with the different parts of the town in which he may be appointed to act, and notice the declivities of the different streets, etc. He will find this knowledge of great advantage."

"...[in examining his fire data from 1824-1829] serious fires decrease as the number of alarms increase... the cause of so many alarms... arise from foul chimneys... the number of houses, shops, and assessable places... is 29,000...average of fires for... five years is about 105... cases of foul chimneys... being one fire to each 276 houses."

Braidwood knew in 1830 that *knowledge of your community* is essential for creating a fire prevention strategy.

Effective Fire Prevention Organizations: *Have a plan.*

James Braidwood wrote, "The person having the principal charge... should frequently turn over in his mind what might be the best plan. By frequently ruminating on the subject, he will find himself much more fit for his task than if he had never considered the matter at all."

Effective Fire Prevention Organizations: *Enforce the code.*

Braidwood writes, "As almost all fires arise from carelessness in one shape or another, it is of the utmost importance that every master of a family should persevere in rigidly enjoining, and enforcing on those under him the necessity of observing the utmost possible care, in preventing such calamities, which, in nineteen cases out of twenty, are the result of remissness or inattention."

Effective Fire Prevention Organizations: *Are proactive with plan review and field inspections.*

Braidwood made the following observations related to the building construction and field inspection process:

"Great carelessness is frequently exhibited by builders, when erecting at one time two or three houses connected by mutual gables, by not carrying up the gables or party-walls with a skew on the outside, so as to divide the roofs."

"It is not uncommon thing, too, to find houses divided only by lath and standard partitions, without a single brick in them."

"The subject of fire-proof buildings might occupy a considerable space. To make a building fire-proof, the stairs must be of stone, and the doors of iron."

"...the next thing to be considered is a supply of water."

Effective Fire Prevention Organizations: *Investigate fire incidents.*

Through the fire origin and cause investigation process, Braidwood was able to determine and track the most frequent causes of fire:

"The most immense hazard is frequently incurred for the most trifling indulgences, and much property is annually destroyed, and valuable lives often lost, because a few thoughtless individuals cannot deny themselves the gratification of reading in bed with a candle beside them."

"...leaving their houses to the care of children."

"Intoxication is also a disgraceful and frequent cause of fire."

"...cinders falling between the joints of the outer and inner hearths."

"...foul chimneys."

Effective Fire Prevention Organizations: *Educate the public.*

Braidwood's public education messages resemble many of today's public education messages:

"When a fire actually takes place, every one should endeavor to be as cool and collected as possible...."

"The moment it is ascertained that fire has actually taken place, notice should be sent to the nearest station where there is a fire-engine."

"...shut all the doors and windows as close as possible, which greatly retards the progress of the flames...."

Effective Fire Prevention Organizations: *Are adequately staffed.*

Tasked with creating this fire organization, Braidwood required and requested eighty men. Citing "budget constraints," his request for eighty men was reduced to fifty men. Eventually, after showing the value of his organization and what his men were capable of, he was given all the personnel he needed. He then established the first shift, station, and company designations and divisions.

"...however complete in its apparatus and equipments, must depend for its efficiency on the state of training and discipline of the firemen. Wherever there is inexperience, want of co-operation, or confusion amongst them, the utmost danger is to be apprehended in the event of fire."

"The description of men from whom I have been in the habit of selecting firemen are slaters, house-carpenters, masons, plumbers, and smiths."

"In each company, there is one captain, one sergeant, four pioneers, and six or eight firemen."

With limited personnel, fire prevention organizations are being forced to function more effectively and efficiently than ever before. Many fire departments have few dedicated fire prevention personnel covering many square feet of space, responsible for the task of life safety inspections, fire protection system testing, plan review, investigations, public education, and additional administrative tasks.

Final Word

With the many tasks, responsibilities, and requirements of the fire prevention organization, how can we best utilize our personnel and ensure that they are functioning most effectively and efficiently by focusing on the right things? It has been my goal in this short title to provide a *Fire Prevention Blueprint*, a tool, guide and resource that you can utilize to increase the effectiveness and efficiency of your organization.

All tools, resources, and references in this book can be found and accessed from this book's resource page at:

www.FPOblueprint.com.

Annex: A Risk Assessment Model

There have been some discussions on methods of establishing definitive risk categories for structures. A uniform method so that what one jurisdiction considers high-risk would not be considered moderate or low risk by others, but a consistent method to evaluate these risks. Below is one example of a risk assessment methodology.

The complete tools, resources, and a book outlining this method can be accessed at www.aviationrisk.blogspot.com.

Process:
1. Tools utilized: Risk Assessment_Field Checklist

Walk through all buildings on the facility, and thoroughly fill out the field checklist information.
Also, review structure for any fire protection/life safety code compliance issues.

    a. Walk-through and fill in notes.
    b. Take any necessary measurements (exposures, hydrants, etc.)
    c. Photograph all four sides of structure (label "A,B,C,D" or "N,S,E,W").
    d. Sketch plot plan (to show building footprint and surrounding
    structures/hazards/hydrants/components).
    e. Sketch a rough floor plan, specifically marking FACP and Riser locations.

2. Tools utilized: digital pre-plan template

Input the information from the field checklist into the corresponding pre-plan template
fields. Attach all photos, drawings, etc.
    a. Fill in information fields.
    b. Attach overhead photo from google earth.
    c. Mark location of hydrants and compass direction on photo.
    d. Attach CAD floor plans (or your drawing).

Fire Prevention Blueprint

3. Tools utilized: Risk Assessment Score Sheet Matrix

Utilizing the field checklist and digital pre-plan information from the previous steps, complete the hazards analysis/risk assessment score sheet matrix.

    a. Identify each building to be evaluated (this goes in the top row).

    b. Evaluate the Potential Hazards (top section of matrix): These hazards are based on three factors: (1)ignition sources present, (2)fuel load present, (3)occupant load. These are evaluated on a scale of 6 – 10 (6 being lowest potential, 10 being highest probability risk). Add up the numbers in these columns, the total is an indication of the potential risk (the greater the number, the greater potential risk).

    c. Evaluate the fire risk reduction factors (bottom section) There are 5 factors considered here:

    (1)building construction type,
    (2)fire alarm system,
    (3)fire suppression system,
    (4)building upgrades,
    (5)water supply and reliability. Each of these can earn up to 2 points, as follows:

        (1) Building Construction Type
            a. Type I, II, or IV = 2 points
            b. Type III, or V = 1 point
        (2) Fire Alarm System
            a. Installed, maintained, and fully functional fire alarm = 2 points
            b. Partial fire alarm, partial protection (i.e, not maintained, waterflow only, etc.) = 1 point
            c. No fire alarm installed = 0 points
        (3) Fire Suppression System
            a. Installed, maintained, fully functional, fire sprinkler/suppression system appropriate to hazard protected = 2 points
            b. Partial system or system not current with all ITM = 1 point

Annex: A Risk Assessment Model

        c. No suppression system in place = 0 points

    (4) Building Upgrades

        a. Significant upgrades to structure or life safety systems (50% or more) within the past 10 years = 2 points

        b. No recent or significant upgrades, but structure in compliance with current codes and standards = 1 point

        c. Other = 0 points

    (5) Water Supply and Reliability

        a. Municipal or other dedicated water supply with hydrants within 250' of any part of structure = 2 points

        b. No dedicated water supply or hydrants greater than 250' to structure = 1 point

        c. Other = 0 points

        d. You will subtract this total number from the potential hazards/risk number, this total will indicate the total potential fire/life safety risk.

Low Risk - 7-15
Medium Risk - 15-21
High Risk - 22-29

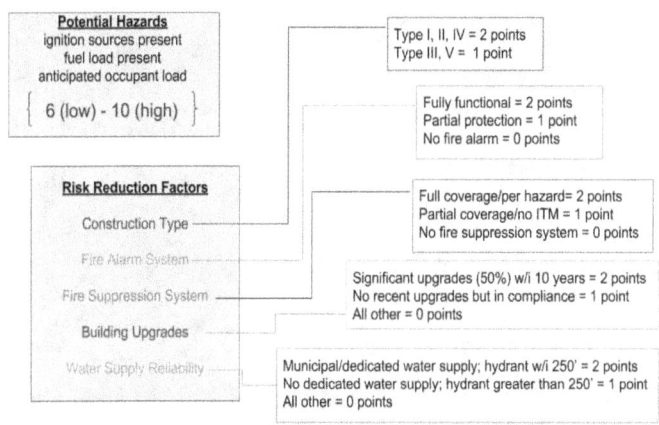

About the Author

Aaron Johnson has more than a decade of fire protection, life safety, and code compliance experience. He is the author of more than 450 industry books, articles, reports, white papers, and blog posts. He holds multiple fire service certifications and is active on multiple technical committees. Aaron resides in South Florida where he is a fire marshal and freelance consultant.

Connect with Aaron:

www.TheCodeCoach.com
LinkedIn - www.linkedin.com/in/baaronj
Twitter - twitter.com/thecodecoach
E-mail - thecodecoach@gmail.com

Other books by Aaron:

Sun Tzu and the Art of Fireground Leadership

Risk Assessment Guide for Aviation Facilities

The Unaccounted Four (fiction)

www.ingramcontent.com/pod-product-compliance
Lightning Source LLC
Chambersburg PA
CBHW030049230526
45471CB00003B/1015